CONTENTS

This symbol shows where there is some information to help you stay safe at the seashore.

Words in **bold** can be found in the glossary on page 30.

THE SEASHORE

On the seashore are many curious things. You can find out about them in this book.

Sometimes you have to guess what you see, then turn the page to find the answer.

Near the end of this book is our seashore curiosity box. You can talk about it with your friends.

You can make your own curiosity box too, if you visit the seashore.

THE TIDES

The seashore is the place where the sea and the land meet.

The sea comes up the shore and then goes back down. In most places it does this twice a day.

As the sea goes up or down we say the tide is coming in or going out.

At high tide the beach is covered by the sea.

At low tide the shoreline is further down the beach.

WAVES

Far out to sea, the wind blows the water into waves. When waves reach the shore they crash onto it.

If the waves are large, they move the sand and pebbles or break pieces off the rocks.

SAND DUNES

You often walk through mounds of sand – called sand dunes – to get to the beach. The dunes hardly ever get covered by the sea.

MARRAM GRASS grows on many dunes. Its **roots** hold the sand together so it does not blow away.

Marram grass grows lots of tiny flowers on one stem. They look like spears.

Sea holly has a thick stem and beautiful green and blue leaves.

SEA HOLLY has spiky leaves to stop animals eating it. Can you spot the dome of flowers at the top of the stem?

The GRAYLING BUTTERFLY is an **insect**. It feeds on the **nectar** of the sea holly and lays its eggs on the marram grass.

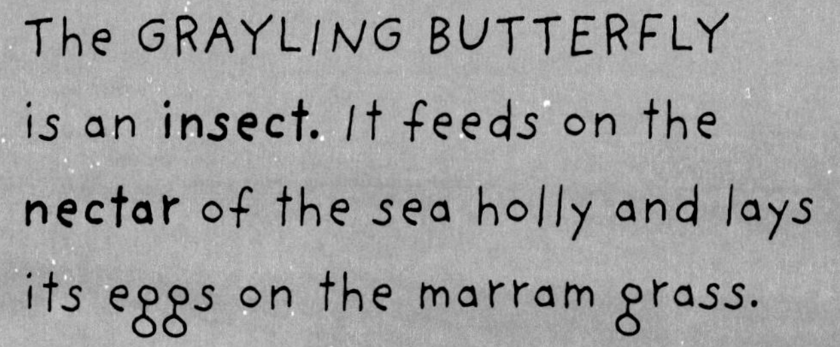

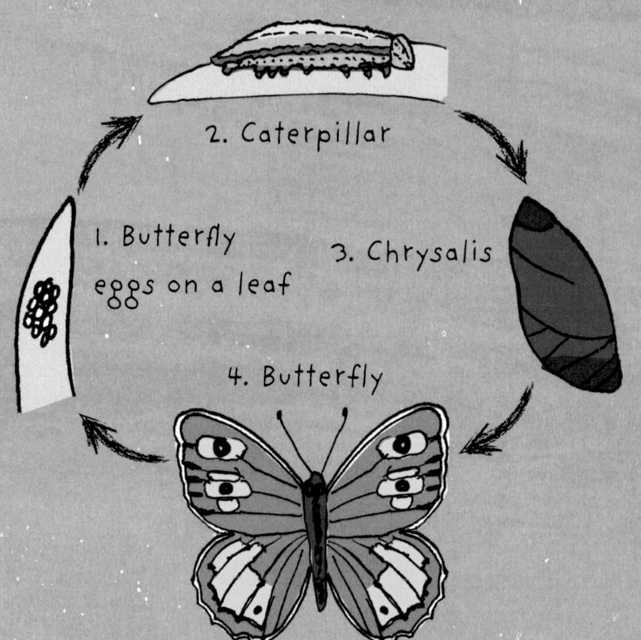

A caterpillar hatches out of each egg, eats the leaves and then turns into a **chrysalis** (*kri-sa-lis*). A new butterfly will emerge from the chrysalis.

WHAT CAN THIS BE?

Is it paper?
Is it plastic?
Is it skin?
Turn the page to find out.

It's sand lizard skin!

Like all **reptiles**, a sand lizard gets rid of its old, dry skin. Its new skin is under the old skin. It lives in a **burrow** in the sand dunes. In winter the lizard **hibernates**.

CRICKETS are insects that live on sand dune plants. The sand lizard eats crickets!

Male crickets rub their wings together to make a chirping sound. This attracts females.

Wild RABBITS make burrows in the sand dunes, too.

SHELDUCKS make nests in old rabbit burrows. They lay up to 14 eggs. When the ducklings hatch, their mother takes them to the sea.

These shelduck ducklings are learning to feed on small sea animals that live in the sand.

THE STRANDLINE

Below the dunes at the top of the beach is the strandline. You will find lots of things here. They have been washed onto the beach by the tide.

Dead SEAWEED makes up most of the strandline. There may also be rubbish made of plastic, glass and metal. You must not touch these things.

This is DRIFTWOOD. It has been in the sea for a long time. The water has worn it smooth.

You may spot lots of tiny holes in a piece of driftwood. They are made by **shipworms**.

A WHELK is an animal called a **mollusc**. It is like a garden snail, but whelks live on rocky seashores instead of in gardens.

A whelk shell is thick and strong to protect the animal that lives inside it.

Whelk egg cases look like plastic when they are stuck together.

Here are the empty egg cases of a whelk. After the baby whelks hatch, the egg cases float away and many wash up on the strandline.

WHAT CAN THIS BE?

Another empty egg case?
A piece of plastic rubbish?
A pirate's eye patch?
Turn the page to find out.

It's the empty egg case of a fish called a skate!

The long 'arms' hold the egg case on the sea floor while a baby fish forms inside.

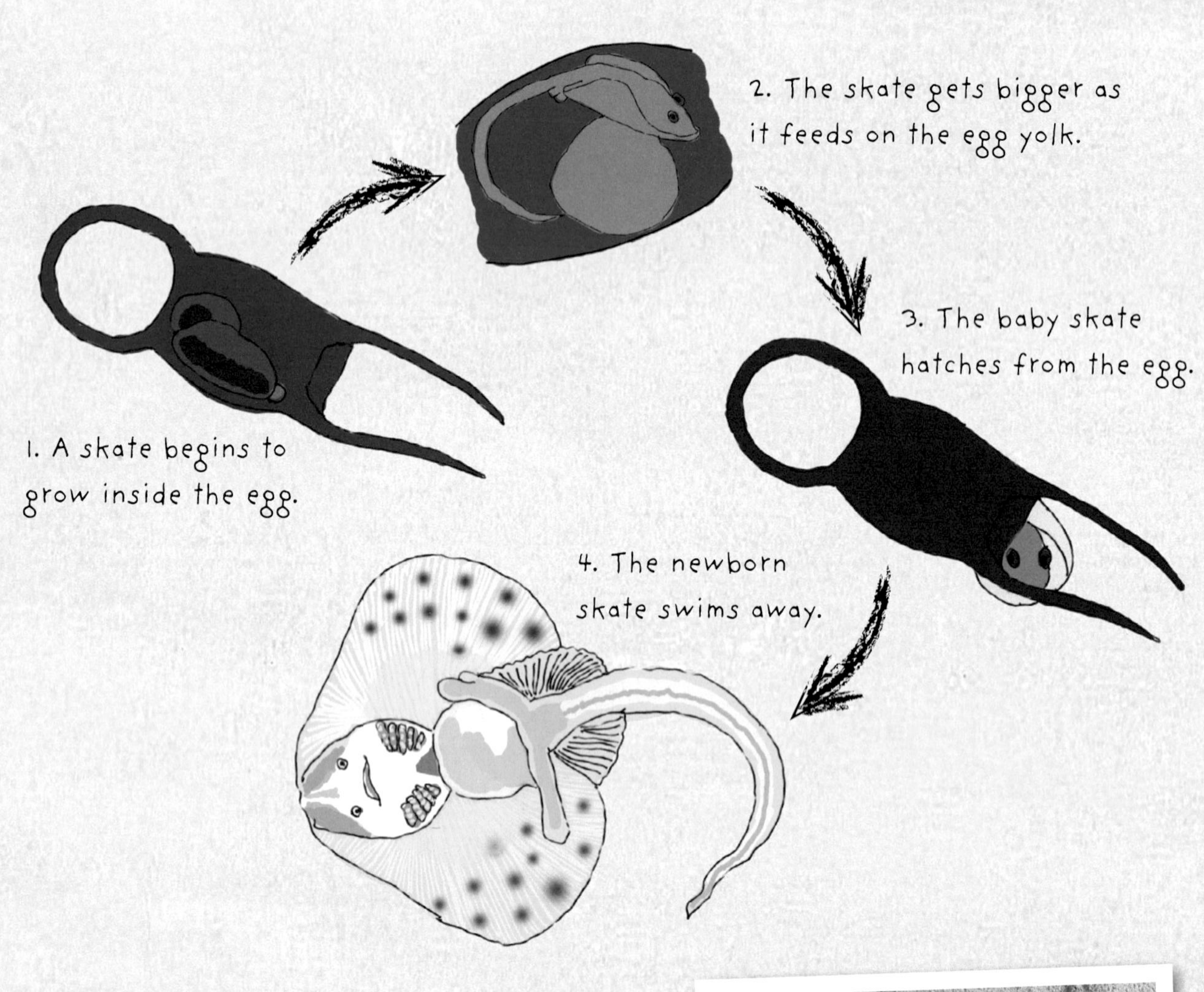

The skate is a flat fish that lives on the sandy sea floor. It feeds on shrimps, crabs and smaller fish.

A skate's flat body helps it burrow under the sand on the sea floor to hide from **predators**.

These strange oval-shaped objects often wash up onto the strandline. It is a CUTTLEFISH bone. It is a kind of shell, but it is found inside a cuttlefish. When a cuttlefish dies, its body rots but the bone does not.

Cuttlefish bones are full of **gas**. This gas helps the cuttlefish float in the water.

Cuttlefish can change the colours and patterns on their bodies.

Cuttlefish are molluscs, which live in groups called shoals. They swim near the shore and catch shrimps, crabs and fish with their **tentacles**.

The tide has left a dead JELLYFISH on the strandline. Jellyfish live in the sea and have long, stinging tentacles to catch fish.

⚠ You must not touch the tentacles because they can sting you even when the jellyfish is dead.

Jellyfish have umbrella-shaped bodies. The jelly is mostly water.

THE SANDY BEACH

When you walk along a sandy beach you may see lots of different empty shells. These shells come from molluscs that live in the sand. Each animal has two shells that are similar shapes.

Here are some of the different kinds of shell you will find on a sandy beach.

The animals that live in these shells are also called **bivalves**. They have a single foot that they use to burrow into the sand.

The RAZOR SHELL uses its foot to dig into the sand. It has two tubes to take water in and out of the shells. The animal takes tiny pieces of food out of the water and eats them.

A razor shell with its foot sticking out of its shells.

A dunlin pulls a sandworm from the beach near the water's edge.

The *DUNLIN* is a **shorebird** with a long, thin **bill** that it pokes into the sand. It eats sandworms, insects and molluscs but it does not eat their shells.

WHAT CAN THIS BE?

The shell of an animal that hides under stones?
The shell of a sand burrower?
The shell of a swimmer?
Turn the page to find out.

The animal is a swimmer that has washed up on the beach!

It is a mollusc called a scallop. It moves by clapping its shells together.

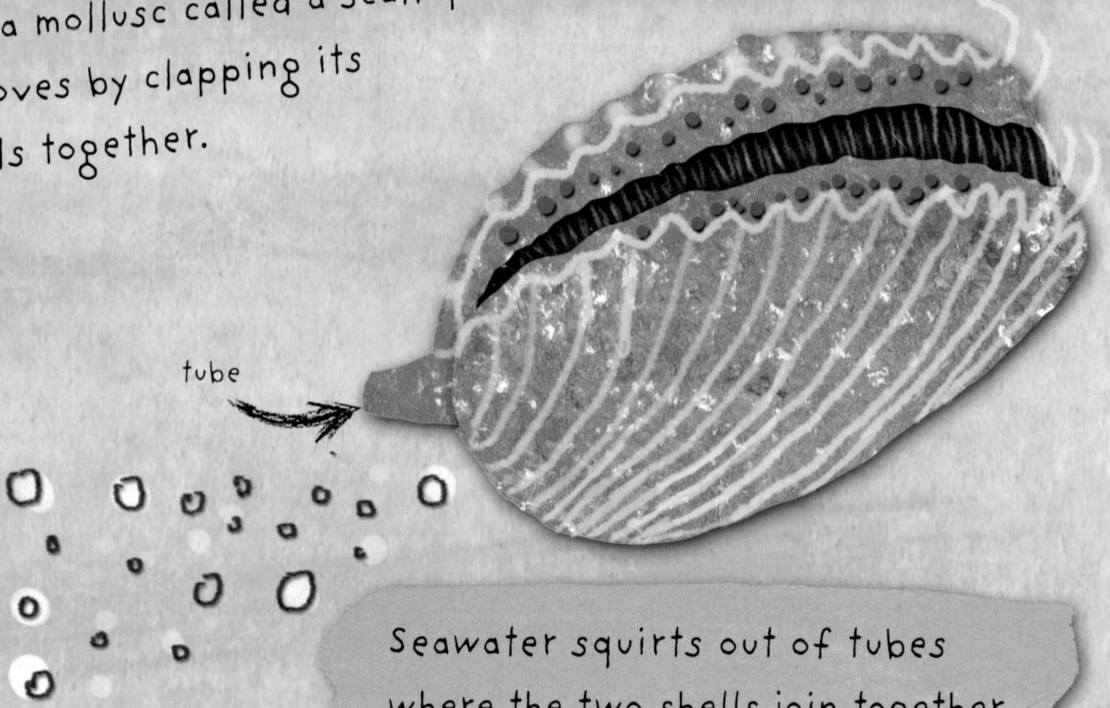

tube

Seawater squirts out of tubes where the two shells join together. This pushes the scallop along.

A scallop has lots of eyes. It has to open its shell a little to see where to go.

eyes

There are different kinds of WORM living in the sand.

The SAND MASON WORM makes a tube from grains of sand and broken shells. The tube sticks up out of the sand.

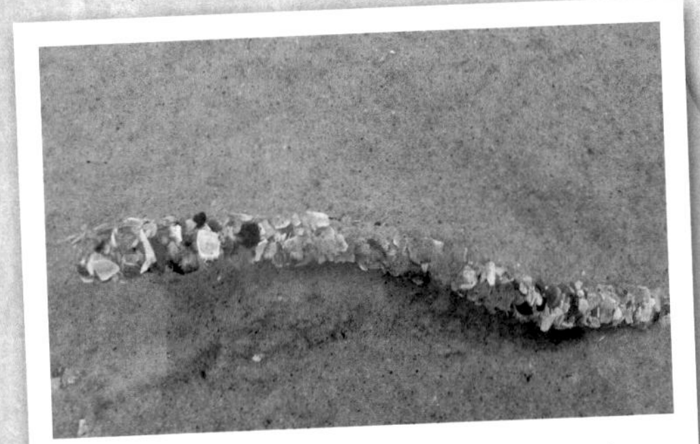

One end of the sand mason's tube is stuck in the sea floor.

The tentacles on a sand mason's head help it to catch food.

When the tide comes in, the worm pops out of the tube and feeds on small bits of food in the water.

The LUGWORM does not make a tube. It lives in a burrow in the sand.

On top of each lugworm burrow you can see a **cast**. This is a coil of sand the lugworm has eaten to find food.

Lugworms are sometimes used as **bait** to help catch fish.

THE ROCKY SHORE

Next to a sandy beach you may find a rocky shore with seaweeds on the rocks. Three kinds of seaweed you find on the rocky shore are bladderwrack, flat wrack and saw wrack.

These seaweeds do not have roots. They have **holdfasts** to grip rocks, so they are not washed away by the tide.

saw wrack

flat wrack

bladderwrack

Here is a large seaweed called KELP. Its holdfast is attached to rocks deep underwater. Sometimes the holdfast breaks and the kelp washes up onto the shore.

Kelp has 'leaves' that look like long, flat blades.

MUSSELS are molluscs that stick to rocks with threads. The threads stop them being washed away. When the tide comes in, they open their shells to feed.

Mussels feed on tiny plants and animals that float in the water.

Some types of limpet move back to the same spot every day when the tide is out.

LIMPETS are a kind of sea snail. They have just one shell. It is shaped like a cone.

Limpets cling to rocks when the tide is out. They feed on seaweed when the tide is in.

WHAT CAN THIS BE?

An animal like a mussel?
An animal like a limpet?
A different kind of animal?
Turn the page to find out.

It's a different kind of animal!

The animal is a barnacle. It sticks to rocks in a different way to mussels or limpets. It is not a mollusc, it is an animal that belongs to the same animal family as crabs.

A baby barnacle sticks itself to a rock with a type of glue and grows a shell with 'doors'. It will stay there all its life.

You can tell that a barnacle has died when you see a barnacle shell without a 'door'.

When the tide comes in, the barnacle opens its doors and sticks out its legs. It waves them around to bring food into its shell.

There are many other animals on the rocks with coiled shells. These are all types of SNAIL.

The cowry is a little different – it is a sea snail with an egg-shaped shell.

ROCK POOL LIFE

When the tide goes out, some seawater is left among the rocks. This is called a rock pool.

Bright green and pale red seaweeds grow in rock pools. Animals such as crabs, shrimps and fish live in rock pools too. They can be hard to spot among the seaweed.

If you look closely you may see a SHRIMP swimming in a rock pool.

Shrimps' colours help them blend in with the rocks in rock pools.

ROCK GUNNELS are eel-like fish that can sometimes be seen at the edge of a rock pool. Most fish breathe underwater through their **gills**, but rock gunnels can also breathe air.

The rock gunnel slithers away quickly if it feels it is in danger.

An anemone's mouth is in the middle of its tentacles.

You may see something that looks like red jelly on the rocks. They are SEA ANEMONES (*an-em-on-ees*). When the tide comes in they pop out lots of tentacles to catch shrimps and little fish. They look a bit like plants, but they are animals.

WHAT CAN THIS BE?

A whelk coming out of its shell?
A crab hiding in a shell?
A piece of seaweed in a shell?
Turn the page to find out.

It's a hermit crab hiding in a shell!

Crabs are animals called **crustaceans** (*crus-tay-shuns*). Most crabs have a strong, hard body, but a hermit crab's is soft. It finds an empty seashell to live in. It uses the shell to protect itself.

As the crab grows, it looks for a bigger shell to live in.

Shore crabs have very strong claws. They can give you a nip!

Lift up the seaweed at the side of a rock pool and you may find a SHORE CRAB. If it waves its claws at you, it wants you to keep away.

Starfish don't just use their feet to move – they breathe through their feet, too!

Echinoderms (*eh-key-no-derms*) are another type of rock pool animal. STARFISH have hundreds of tiny feet under their arms, but they move... very... slowly. If you watch one for a long time, you may see it move.

Most starfish have five arms. If they lose an arm, they can grow another one!

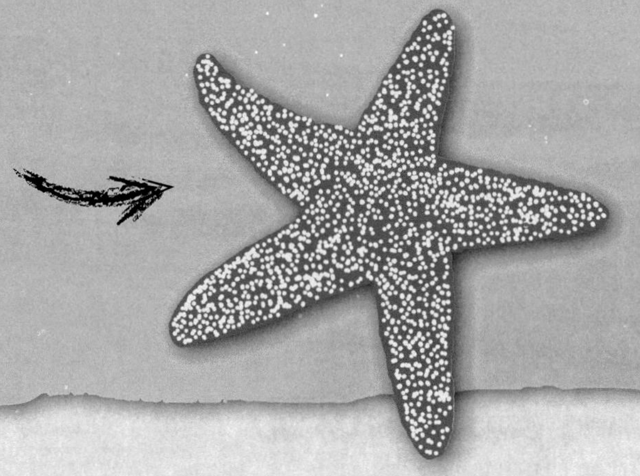

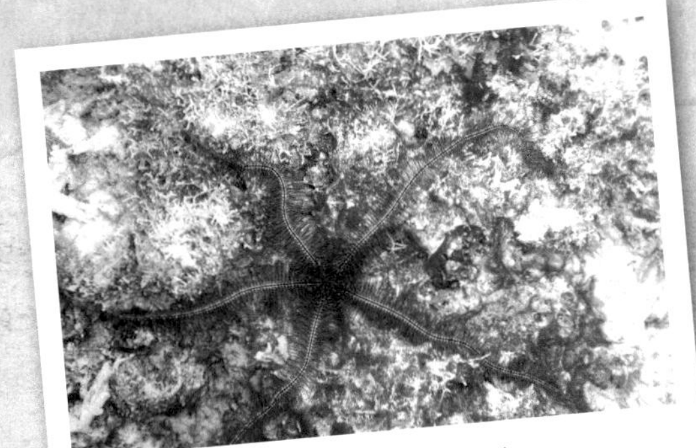

The brittle star has five arms too – just like the starfish.

If you look under a stone you may see a BRITTLE STAR. It swims by waving its arms about.

Or you may see a SEA URCHIN feeding on seaweed. It has spiky skin that covers its round body.

You must not pick up sea urchins as their spines can sting you.

Sea urchins used to be called sea hedgehogs!

THE CLIFFS

Above the rock pools, you may see some tall rocky cliffs. The sea crashes into the rocks at the base of the cliff.

Lots of SEABIRDS nest on the cliffs. High up here their eggs are safe from predators.

This white-and-grey seabird is called a kittiwake.

These black-and-white seabirds are called razorbills.

You can find feathers at the bottom of many cliffs. These feathers are from a kittiwake's wing.

Below the birds' nests you may find a SKELETON of a fish that has been dropped by a seabird. Can you tell which end is the tail?

Most of this fish has been eaten. Just the head, backbone and tail are left!

A big **mammal** is lying on the rocks at the bottom of this cliff. It has flippers, big black eyes and it eats fish. What do you think it is?

SEASHORE CURIOSITY BOX

A curiosity box is a place to put all of the curious things you have collected.

What items are in your seashore curiosity box?

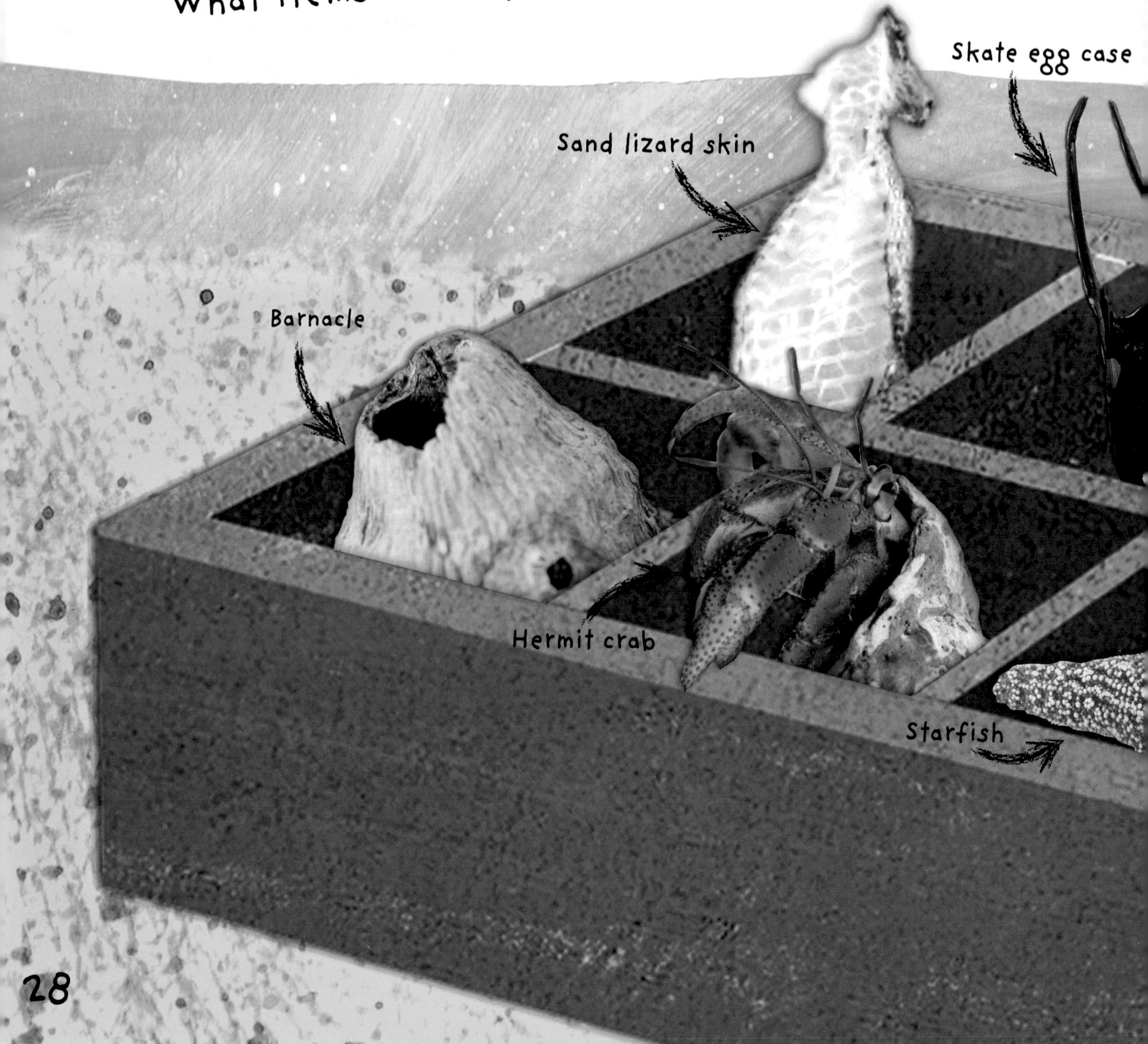

CURIOUS QUIZ

1. Where do you find marram grass?
a) on a sand dune
b) in a rock pool
c) on a cliff

2. Which animal has two long, narrow shells?
a) a common cockle
b) a razor shell
c) a scallop

3. Which of these is a type of seaweed?
a) fat wrack
b) flat wrack
c) float wrack

4. Which animal has a coiled shell?
a) a limpet
b) a cuttlefish
c) a tower shell

5. Where would you find a sea anemone?
a) in a rock pool
b) on a sandy beach
c) on the strandline

6. Which bird nests on cliffs?
a) a shelduck
b) a dunlin
c) a kittiwake?

Curious quiz answers: 1a; 2b; 3b; 4c; 5a; 6c.
The animal on the rocks on page 27 is a SEAL!

GLOSSARY

bait food that people put on hooks to catch fish

bill another name for a bird's beak

bivalves animals called molluscs that have soft bodies and two shells

burrow a type of animal home

cast a coil of sand on the beach

chrysalis the stage in an insect's life when it is inside a hard case as it changes into an adult

crustacean a sea animal with legs and a hard shell, such as a crab or shrimp

echinoderm an animal with a round or star-shaped body and tough skin

gas a material, like air, which is not solid or liquid

gills the parts of the body that let animals living underwater breathe

hibernate to spend winter asleep

holdfast a part of seaweed that looks like a root and holds it onto rocks

insect an animal with a hard body, six legs and usually two or four wings

mammal an animal with a skeleton and hair or fur. They feed on their mothers' milk to help them grow

mollusc an animal with a soft body, such as a snail. Most molluscs have a shell on the outside of their bodies

nectar a sugary liquid that plants produce to attract insects

predators animals that hunt and eat other animals

reptiles animals that breathe air and have dry skin covered in scales

roots the underground part of a plant that takes up water from the soil

shipworms a mollusc that eats wood

shorebird a bird that eats animals living on a sandy or rocky shore

tentacles long, bendy arms of an animal

INDEX

CURIOUS FACTS

CURIOUS BEGINNINGS

People have collected objects for thousands of years. During the 1500s and 1600s, special cabinets were made to display the objects that were brought back from voyages to newly-discovered lands, such as North America. These cabinets were sometimes whole rooms, which became the first museums.

WHAT IS A CURIOSITY BOX?

A curiosity box is a small copy of these cabinets. It is a more scientific way of displaying items than a nature table. You can group items together by season or by theme. For example you can create: a box of seashells; a box of seaweeds; or a box of animal parts, such as lizard skin or feathers.

YOUR CURIOSITY BOX

It's easy to make your own curiosity box. A shoebox or other small cardboard box will do! Ask an adult to help you cut long strips of card with slits cut into them. Slot them together to make lots of small sections inside your box. Place the objects you find (or photographs of them) inside the sections.

USEFUL INFORMATION & WEBLINKS

Many natural items won't last forever. For help with making your curiosities last a bit longer, try websites such as:
www.seashells.org
If you collect shells, select single items only. Make sure they are clean and do not have any flesh attached to them.
For general information on practical science involving plants and animals, contact the Association for Science Education at www.ase.org.uk for their book, *Be Safe!* (Fourth edition).

SEASHORE NOTES

Here is some more information, for parents and teachers, on the animals, objects and habitats found in this book.

The seashore and tides

Sandy beaches are made by waves breaking rocks into smaller and smaller pieces until they are grains of sand. Shingle beaches are made of small, smooth stones. Sandy beaches have more animal and plant life than shingle beaches, due to the constant churning of the stones as the tide comes in and out. Like most shores, the shingle beach does have a strandline. Ask a child to compare shingle with ordinary rocks from a garden to see how erosion wears rocks into smooth shapes.

Tides are caused by the pull of gravity from the Sun and the Moon. When there is a new or full Moon, the Earth, Sun and Moon have formed a straight line and the pull of gravity is at its strongest and so the tides are at their highest.

Sand dunes

Sand dunes can be exciting habitats for children to visit. Wind and waves have moved the sand into large piles, which are often safe and fun to explore. Look for warning signs for any areas that might be dangerous. As you move further inland, observe how the amount of plant life increases, the further you are from the beach.

Marram grass roots help hold the sand together, so the sand dune is stable. This allows other plants to grow on the dunes and further inland it is stable enough for bushes and trees, which in turn support even more animal life.

The strandline

Objects that were once in the sea make up the strandline. Waves wash the objects onto the beach. Most of the strandline is seaweed and becomes a habitat for sand hoppers. These animals are related to woodlice and shrimps.

You may see holes in pieces of driftwood. These holes are made by a mollusc called the shipworm, which bores into the wood and digests the cellulose in the wood.

Cuttlefish are molluscs. You can find cuttlefish bones in pet shops. The bone is made of a mineral called aragonite, which is full of calcium. Pet birds eat the bone to get more calcium into their diet.

The sandy beach

Most of the bivalve shells you find on a beach will be empty as the animal has been eaten or died. Bivalves have a foot that they use to dig down into the sand. As they move the sand, water flows between the grains. This makes digging easier.

Bivalves breathe through gills linked to tubes called siphons. The gills have tiny hairs that wave about and draw water over them. One tube brings water in and the other tube takes water out. The tiny hairs also move food towards the mouth.

The rocky shore

Brown seaweed can be seen at low tide. Red and green seaweeds are found mainly in rock pools. Many types of seaweed are edible and are an important food source around the world. Laver is a common type and can be bought from a local fishmonger if you want to try it.

Barnacles look like molluscs, but are actually related to crabs and shrimps. After hatching, the larval stage swims in the water, then sticks its head on a rock (or even on a boat or a whale) and grows the many crusty plates of its shell around it. When it feeds, it sticks its legs out of its shell to draw water in towards its mouth.

Rock pools

Many rock pool animals, such as shrimp, are camouflaged to match the rock pool floor. They are easier to spot when they are on the move.

Sea anemones are related to jellyfish. They have stinging tentacles around their mouths to paralyse and capture prey. When the prey is still, the tentacles push the prey into the anemone's mouth.

Cliffs

Birds that nest on cliffs lay eggs that are shaped like a rounded cone. They roll around in a circle instead of rolling off the edge of the cliff. Collecting birds' eggs is illegal and children should be taught never to handle wild bird eggs or disturb birds' nests.

Seals use rocks at the bottom of cliffs as places to rest out of the water. A group of seals is called a colony and they can be very noisy and smelly! Orkney in Scotland is one of the best places to watch seals. Children should be encouraged to observe them from a safe distance, as these large animals can be aggressive during the mating season.